TIME TRAVEL

Is it possible?

Tom Held

I0426974

Introduction: What is time travel and how is it possible?

Time travel is a fascinating concept that has long captured the imagination of scientists, authors and filmmakers alike. It's the idea that you can travel into the past or future to change events or influence destiny. But what is time travel and how is it possible? In this essay, we will address these questions and try to get a clear idea of what time travel is and how it might work. The term time travel refers to the idea that you can move from one point in the space-time continuum to another without following the normal flow of time. In other words, time travel would mean having the ability to manipulate time and being able to travel between different points in time as if you were immersed in another dimension. The idea of time travel has long been a staple of science fiction literature and film, but it has also generated a great deal of interest in the scientific community. There are numerous theories about how time travel could be possible, but to date there is no definitive answer to this question. One of the best-known theories is the theory of relativity developed by Albert Einstein. The theory of relativity states that space and time are inextricably linked and that the curvature of space affects the curvature of time. This means that time can pass slower or faster depending on how strong the gravitational force is in a particular area. The theory of relativity has shown that time travel could, in principle, be possible if a curvature can be created in the space-time continuum that allows an object to move faster than light. Another theory being discussed in the scientific community is quantum mechanics. Quantum mechanics is the study of subatomic particles and their behavior. It states that particles are connected due to quantum entanglement and that changes to one particle can have an effect on another particle, even if they are spatially separated. Some scientists believe that it may be possible to use this connection to travel back in time. There is also a theory called the Closed Timelike Curve (CTC), which states that it may be possible to create a kind of space-time continuum that allows an object to travel back in time and change events. This theory is based on the idea that the curvature of space and time could be so strong in certain cases that it would be possible to find a way to travel through time.

Time travel in literature and film: a historical overview

The idea of traveling through time has fascinated people for a long time. It is a concept that is widely used in literature and film. In this essay, we will look at the historical development of time travel in literature and film. The beginnings of time travel in literature can be traced back to H.G. Wells' novel "The Time Machine" from 1895. In this novel, the protagonist travels into the future, where he discovers a dystopian world. This novel is considered one of the first examples of time travel in literature and had a major influence on the later development of the genre. In the 1920s, science fiction literature began to focus on time travel. One of the most famous novels of this period is "The Man Who Folded Himself" by David Gerrold from 1973, in which the protagonist travels through time and is confronted with different versions of himself. In the 1940s, filmmakers also began to take an interest in time travel. One of the most famous films from this period is "A Star Rises" from 1941, in which a young actress travels back in time to further her career. This movie was a huge success and helped make time travel a popular theme in film. In the 1960s, time travel became increasingly popular in literature and film. One of the most famous films from this period is "Back to the Future" from 1985, in which the protagonist travels back in time and must ensure that his parents fall in love so that he himself can exist. This movie was a huge success and became a classic of the science fiction genre. In the 1990s, time travel became an important theme in pop culture. One of the most famous films of this period is "Terminator 2: Judgment Day" from 1991, in which a cyborg from the future travels to the present to protect the young John Connor. This movie was a huge success and led to several sequels. In the 2000s, the trend towards time travel continued. One of the most famous films from this period is "Source Code" from 2011, in which the protagonist travels to a parallel world to prevent a terrorist attack. This film was a critical and commercial success and helped to further popularize time travel in pop culture. In recent years, interest in time travel in pop culture has not waned. One of the most popular films in recent years is 2019's Avengers: Endgame, in which the superheroes of the Marvel Cinematic Universe travel back in time to defeat 2014's Thanos. This movie was a huge commercial success and broke several records.

The physics of time: time travel and relativity

Time is one of the most fundamental concepts in physics and one of the most important dimensions that influence our understanding of the world around us. However, the human perception of time is very subjective as it is influenced by many factors, such as individual experience, environment, cultural traditions and the physiology of the brain. In physics, however, time is an objective quantity that can be measured and quantified. In this context, the theory of relativity and quantum physics have fundamentally changed our understanding of time. The theory of relativity One of the biggest influences on our understanding of time is the theory of relativity, which was developed by Albert Einstein in 1905. This theory describes how space and time change relative to each other based on the speed of objects. The theory of relativity has two main parts: special relativity and general relativity. The special theory of relativity (SRT) describes the relationship between space and time in a relativistic space-time continuum. The theory states that space and time are inextricably linked and that time cannot be regarded as an absolute quantity. Time dilation is an important consequence of SRT, which states that time passes more slowly when an object is moving at high speed. This effect has been confirmed by experiments such as the famous twin paradox. The general theory of relativity (ART) describes the relationship between gravity and space-time. ART states that gravity is a property of space and that masses and energy influence the curvature of space. The curvature of space in turn influences the movement of objects in space and therefore also time dilation. An example of this is time dilation in the vicinity of massive objects such as black holes. Time travel The idea of time travel has become very popular in science fiction literature and films. The ability to travel into the past or future would open up the possibility of changing events or influencing future events. In physics, however, time travel is a theoretical possibility and there are various theories about how it could be possible. One possibility of time travel is the so-called "time loop". This is a concept that states that time runs in a loop and that it is possible to break a loop to travel to the past or future. An example of this is the "grandfather paradox", where someone travels back in time and tries to kill his grandfather before he has fathered his father.

The time jump

How far back or forward in time can we jump? The concept of time-jumping, or the ability to jump backwards or forwards in time, is a fascinating topic that has been explored in numerous science fiction stories and films. But how realistic is such a time jump? And how far back or forward in time could we theoretically jump? First of all, it is important to understand that the idea of a leap in time does not really exist in science. To this day, time travel is a purely hypothetical concept based on Albert Einstein's theory of general relativity. This states that space and time are inextricably linked and that gravity bends the fabric of space and time. Some theories suggest that it may be possible to use this curvature to travel through time. However, there are some theoretical limits to the ability to travel in time. One of these limits is causality, or the principle of cause and effect. If we travel back in time and perform an action that changes the past, this could lead to a paradox. For example, by killing your own grandfather in the past, you could prevent yourself from ever being born. However, this would lead to a paradox as you would never have existed to kill your grandfather. Another problem with the idea of time travel is the energy required for such a jump. The theories state that an enormous amount of energy is required to create the curvature of space and time necessary for time travel. However, there is no known source of energy sufficient to create such a curvature. Despite these theoretical limitations, there are some suggestions as to how time travel could possibly be realized. One possibility is to create a so-called wormhole connection between two points in space and time. A wormhole is a hypothetical connection between two areas in space that can be traversed faster than light. Theoretically, a wormhole could also serve as a connection between two different points in space and time. Another possibility is to distort space-time by reaching extremely high speeds and thus travel through time. Albert Einstein's theory of special relativity states that time passes more slowly if you move at higher speeds. So if you were to fly in a spaceship at close to the speed of light, time would pass more slowly than it does on Earth. This could theoretically be used to travel into the future. However, both of these possibilities are still purely hypothetical and it is unclear whether they can ever be realized.

Time travel and paradoxes

Is it possible to change the past? Time travel and paradoxes have always been a fascinating topic in science and entertainment. Movies, books and TV series have fired people's imaginations and raised the question of whether it is possible to change the past. In this essay, we will look at the theory of time travel and the paradoxes associated with this idea, and examine whether it is actually possible to change the past. The theory of time travel states that it is possible to travel through time, either forwards or backwards, and thus influence events of the past or the future. There are various theories that explain how time travel could be possible, including Albert Einstein's theory of general relativity and quantum mechanics. The theory of general relativity states that space and time are inextricably linked and that gravity causes the curvature of space and time. Quantum mechanics, on the other hand, states that the world is made up of unpredictable events known as quanta. However, these theories do not provide a clear answer as to whether time travel is actually possible. However, if we assume that time travel is possible, the question arises as to whether it is possible to change the past. There are several theories that address this question, including the causal loop theory and the theory of parallel universes. The causal loop theory states that the past is unchangeable and that anything we do to try to change it will in fact cause events to unfold in such a way that we do exactly what is necessary to make the past the way we experienced it. This is called the principle of self-consistency. For example, if we travel back in time and try to change an event, our intervention could actually cause the event to occur exactly as it should in order to enable our presence. The theory of parallel universes states that for every decision we make, there is a parallel universe in which we made a different decision. So if we travel back in time and try to change an event, we would indeed create a parallel universe where the event is different. In our own universe, however, the past would remain unchanged. The idea of the causal loop and parallel universes may seem contradictory at first glance, but they both offer a possible solution to the paradox of the changeable past. Paradoxical events, in which a change in the past leads to events in the future no longer being possible, can thus be avoided.

Time travel and determinism

Is the future already predetermined? The idea of time travel is one of the most fascinating and speculative concepts we as humans have ever invented. The possibility of traveling through space and time and visiting the past or the future is a staple theme in science fiction literature, movies and series. However, this discussion is not about the feasibility of time travel, but about a deeper question: is the future already predetermined? This question is often posed in the context of determinism, a philosophical position that holds that all events and actions are determined by prior causes. This idea assumes that the future is already determined, as all events and actions in the past and present predetermine the course of events in a certain direction. The idea of determinism is at odds with the idea that we as humans are free and can make our decisions independently of external factors. If the future is already determined, do we really have the freedom to make decisions? Or are we just puppets following the strings of fate? A common argument against determinism is the concept of human freedom. We feel that we can make decisions that affect our lives and that we are able to choose our own path. If the future is already predetermined, this means that our choices have no impact and that our lives are just going along a predetermined route. Another argument against determinism is the unpredictability of the world. The universe is so complex that even small changes in the initial conditions can lead to completely different outcomes. For example, a butterfly flapping its wings in Brazil could cause it to rain in Japan. This is known as the butterfly effect and shows that even minor changes in the present can radically alter the course of events in the future. However, there are also arguments in favor of determinism. One argument is that all events and actions are linked by causes and effects. When we perform an action, there is always a cause for it, be it a decision we have made or an external influence. This suggests that the future is already predetermined, as it is determined by actions and decisions in the past and present. Another argument in favor of determinism is that many scientific theories suggest that the future is predetermined. Quantum mechanics, for example, states that the states of particles are determined by probability distributions, but these probabilities are deterministic and are determined by the laws of nature.

Time travel and multiverses

Are alternative timelines possible? Time travel and multiverses are fascinating concepts that are often discussed in science fiction literature and movies. Many of us have imagined what it would be like to travel into the past or future to change certain events or influence destiny. But is this even possible? Are alternative timelines realistic or are they just a figment of our imagination? First of all, we need to understand what time travel and multiverses are. Time travel refers to the idea that you can travel back or forward in time to change or observe events. Multiverses are the idea that there are an infinite number of other parallel universes besides our universe in which every possible combination of events takes place. The possibility of time travel is considered possible in principle by physics. According to the general theory of relativity, time is part of the space-time continuum and can be bent by the presence of masses or energy. This means that it is theoretically possible to travel in time by using space-time distortions such as the influence of gravitational fields or by using wormholes.

However, there are some technical challenges and problems that need to be solved before we can put time travel into practice. One of the main problems is the issue of causality. If we travel back in time and change the past, this could have effects on the present and the future that we cannot foresee. This could lead to paradoxes, such as the so-called grandfather paradox, where changing the past means that you yourself are not born. The concept of multiverses, on the other hand, is controversial in physics. Although there are some theories that predict the existence of multiverses, there is currently no experimental evidence for this. One such theory is the so-called "many worlds interpretation" of quantum mechanics, which states that for every quantum event there is an infinite number of possible outcomes, and that each possibility exists in its own universe. So when we bring together the idea of time travel and multiverses, interesting questions arise. If we travel back in time and change the past, would we create an alternate universe? Would we live a different life in this universe because our actions in the past have led to different outcomes? The answer to these questions depends on which theory of multiverses you accept. If one accepts the "many worlds" interpretation, then there are already an infinite number of alternative universes in which every possible combination of events takes place.

Time travel and ethics

Should we intervene in the past? The idea of time travel has always fascinated mankind. What if we could go back in time and change events? Would we be able to influence or even improve the course of history? But what about the impact on the future? The possibility of time travel raises many ethical questions, in particular whether or not we should intervene in the past. First of all, there are a number of practical concerns that argue against intervening in the past. If we go back in time, we could change events that have an impact on the future. Even small changes could set off a chain reaction that ultimately leads to a completely different outcome. For example, preventing the Second World War might seem like a positive action in the short term, but it could lead to an even worse global crisis in the long term. Another practical aspect is the question of how exactly we would intervene in the past. Can we simply travel back in time and change events at will? Or are there certain restrictions or conditions that we need to consider? If we change the past, what effect does this have on our own existence? Would we ourselves still exist or would we find ourselves in a completely different reality? Furthermore, there are also moral concerns that speak against intervening in the past. When we intervene in the past, we intervene in the decisions and lives of other people. We deprive them of the opportunity to make their own decisions and gain their own experiences. We could also lose the opportunity to learn from mistakes and grow as a society. Another moral aspect is the question of who would decide what should and should not be changed. Would we change the course of history according to our own ideas and interests or would we be guided by an objective truth? Who would take responsibility for the changes we make? Despite these concerns, there are also arguments in favor of intervening in the past. One possibility would be to make changes that help specific individuals whose lives have been negatively impacted by historical events. For example, we could try to prevent an assassination attempt on an important political leader that may have triggered a war or other catastrophe. Another argument for intervening in the past could be that we could improve the future. By intervening in the past, we could potentially influence developments that lead to positive outcomes in the future.

Time travel and responsibility

Who bears the consequences of time travel? Time travel is a fascinating topic in science and science fiction. The idea of traveling back or forward in time and changing events has captured the imagination of generations of people. But when we talk about time travel, the question inevitably arises: who is responsible for the consequences? Science fiction often depicts time travelers altering events in the past, setting off a chain reaction of events that dramatically change the future. But who is responsible for these changes? The time traveler himself? Or the society that allowed him to go back in time? One way to answer this question is to refer to the idea of the butterfly effect. The butterfly effect states that small changes in a system can lead to big changes. In terms of time travel, this means that even small changes in the past can lead to big changes in the future. So if we assume that the butterfly effect is indeed applicable to time travel, this means that the time traveler must bear the responsibility for the consequences. Because if the time traveler makes a small change in the past, this can lead to major changes in the future that he cannot foresee. This could lead to catastrophes and suffering for which the time traveler must bear responsibility. Another way to look at responsibility for the consequences of time travel is to refer to the idea of determinism. Determinism states that all events in the world are determined by previous events. So if we assume that time travel is possible, this means that the events that the time traveler changes in the past were already predetermined. In this case, the time traveler would bear no responsibility for the consequences of his actions, as these were already predetermined. However, the question would still remain as to whether it is morally justifiable to travel into the past and change events that have already taken place. A third way of looking at the responsibility for the consequences of time travel is to refer to the idea of multi-world theory. Multi-world theory states that every time a decision is made, the universe splits, creating two parallel worlds in which each decision is made. So if we assume that time travel is possible, this means that the time traveler travels to the past and makes a decision that leads to a new world.

Time travel and identity

How does time travel change our identity? Time travel is a fascinating topic that is taken up in many science fiction stories. It's about traveling to the past or future to change events or simply to see what it looks like in another time. But what happens to our identity when we travel through time? In this essay, I will take a closer look at this question and try to examine possible effects on our identity. The first thing to note is that the idea of time travel is often associated with the idea of change. Traveling to the past and changing something there can have an impact on the future. These effects can in turn have an impact on our identity. For example, if we travel back in time and fix a mistake that we regret in the present, it could change our identity by no longer feeling responsible for that mistake. Another aspect that could influence our identity through time travel is the experience of different times and cultures. When we travel to another time, we may no longer be able to identify with our usual norms and values. We might find ourselves in an environment that is alien to us and forces us to adapt. This could affect our identity by making us rethink and possibly change our beliefs and values. Another interesting aspect is the question of whether we retain our identity when we travel back in time. If we were to travel back in time and meet our younger version, would we still feel like the same person? Or would we see ourselves as strangers? This question is particularly interesting because it also shows how our identity has changed over time. If we would consider ourselves as strangers, this would mean that our identity has undergone changes over time. Another important aspect of considering the impact of time travel on our identity is the question of the freedom and responsibility that comes with it. When we travel into the past and make a decision that has an impact on the future, do we bear responsibility for that impact? Or are we free of this responsibility because we have acted in the past and are therefore no longer responsible for our actions? This is a difficult question to answer, as it is a moral question that depends on many factors. Finally, there is the question of whether time travel is even possible and what impact this would have on our identity. If time travel were indeed possible, it would mean that we could influence ourselves in the past and the future. This would possibly mean that our identity would no longer be as stable as before, as we would find ourselves in different times.

Time as the fourth dimension

Time travel and the geometry of space. The idea of time as the fourth dimension is one of the most fascinating theories in physics and has shaped the idea of time travel and the geometry of space. In this article, we will take a closer look at this theory and examine its implications for our understanding of space and time. The basic idea of time as a fourth dimension goes back to the famous physicist Hermann Minkowski, who first introduced it in 1908 in a speech to the Society of German Natural Scientists and Physicians. Minkowski argued that space and time are inextricably linked and that time is a dimension that is just as important as the three spatial dimensions. Minkowski's idea was later taken up and further developed by Albert Einstein in his special theory of relativity. In special relativity, time is regarded as one of four dimensions that together form the spacetime continuum. This concept states that space and time cannot be considered separately, but that they form an inseparable unit. The idea of spacetime has interesting implications for our notion of time travel. If time is a dimension that is just as important as the spatial dimensions, then it is theoretically possible to travel through time, just as it is possible to travel through space. This has led to a plethora of science fiction stories and movies dealing with the possibility of time travel. One example of such a story is H.G. Wells' famous novel "The Time Machine", which was published in 1895. In this novel, the protagonist travels into the future and experiences a world that has been altered by time. This story shows how the idea of time travel has influenced human thinking and expanded our conception of the world around us. The idea of time as a fourth dimension also has an impact on our idea of the geometry of space. In traditional geometry, objects are viewed in a three-dimensional space defined by the three spatial dimensions. In spacetime geometry, however, space is defined by four dimensions, including time. This extension of geometry has allowed us to view and understand the world around us differently. One example of this is the theory of curved spacetime, which states that the presence of matter and energy can affect space and time and lead to a curvature of spacetime. This concept was introduced by Einstein in his general theory of relativity and has helped to deepen our understanding of gravity and the universe. The idea of curved spacetime shows how the notion of time as a fourth dimension can help to expand our understanding of the world around us.

Time travel and the time machine

How does a time machine work? The concept of a time machine has fascinated and inspired the human imagination for centuries. Movies, books and even scientific papers have explored the possibility of time travel. But how does a time machine work? The idea of a time machine is that it allows you to travel through time, whereby you are able to end up in the past or the future. There are various theories that attempt to explain how such a machine could work, but none have yet been practically realized. One way to build a time machine would be a device that allows you to travel at faster-than-light speeds. This theory stems from Albert Einstein's special theory of relativity, which states that if you travel faster than light, you would be able to go back in time. However, there is a problem with this theory: physics forbids traveling faster than light. It is therefore unlikely that such a machine could ever be built. Another theory suggests that a time machine could be built by using a black hole. A black hole is an object in space that is so heavy that it swallows up everything in its vicinity, including light. If you were to get close to a black hole, the gravitational force would be so strong that it would distort time around you. This means that when you are near the black hole, time passes more slowly than anywhere else in the universe. If you travel fast enough through this slowed-down time, you might be able to arrive in the future without having actually moved. However, there is a problem here too: a black hole is very dangerous and it is unclear whether you could even get close to it without falling into its gravitational pull and being swallowed up. Another theory is that a time machine could be built by using a kind of wormhole. A wormhole is a hypothetical connection between two points in the space-time continuum. If you were to enter a wormhole, you could quickly travel from one point in the universe to another without actually having to cross the space in between. Some scientists believe that it might be possible to artificially create a wormhole to use as a time machine. However, there are some problems here too: it's unclear how you could artificially create a wormhole, and even if you could, it's not certain whether it would be stable enough to allow travel through time. All in all, there are many different theories about how a time machine could work. However, none of these theories are practically feasible, and it is unclear whether it will ever be possible to build a time machine.

Time travel as an experiment

What can we learn from time travel? The idea of traveling through time has long fascinated the human imagination. The idea of traveling to the past or the future to find out what happened or what will happen is a common theme in science fiction literature and movies. But what if time travel was actually possible? What impact would this have on our understanding of space, time and history? And what could we learn from time travel? The concept of time travel is controversial in science as it violates some of the basic principles of physics, in particular causality, which states that cause and effect always run in a certain direction. Nevertheless, there are some theories that allow for the possibility of time travel, such as Albert Einstein's theory of general relativity and quantum mechanics. These theories argue that it is possible to travel through space and time by bending space or moving into other dimensions. So if we assume that time travel is indeed possible, what could we learn from such experiments? One of the most obvious applications of time travel would be the ability to explore the past. We could travel back and observe first-hand events that were previously only known from records or stories. We could experience historical events such as the founding of the USA, the Second World War or the French Revolution and better understand how they actually happened. This would help us to improve our history and our understanding of how societies and cultures came into being. Furthermore, time travel would also allow us to explore the future and foresee possible scenarios and developments. We could see what technologies and innovations will emerge in the next few years or decades and how they will affect our lives and our society. We could also run through different scenarios and see how events would develop if we make or don't make certain decisions. This would help us to make better decisions and shape our future. Another important application of time travel would be the ability to correct past mistakes. We could travel back and prevent disasters such as climate change or the Chernobyl nuclear disaster. We could also correct personal mistakes or mistakes in history by undoing certain decisions or choosing other options. This would help us to grow and develop as individuals and as a society. However, there would also be risks and challenges in conducting time travel experiments. One of the biggest risks would be the so-called "time travel paradox", where a person traveling through time changes the past in such a way that they themselves were never born.

Time travel and causality

How are cause and effect related in time? Time travel and causality are closely linked. In science and philosophy, causality is defined as the relationship between cause and effect. Time travel, on the other hand, refers to the idea that one could travel back in time or into the future. There are many theories and concepts that deal with the question of how cause and effect are related in time. One theory that deals with time travel and causality is the so-called "Novikov self-consistency hypothesis". This states that time travelers cannot do anything to change the past. Instead, every action that a time traveler performs would become part of the past and cause the events that led to them traveling back in time. This concept is based on the assumption that cause and effect are linked and that every action a time traveler takes is already part of the chain of causes that led to the present situation. So if a time traveler travels back in time and tries to change events, they would really only be contributing to those events taking place in a certain way, as their action is already part of the chain of causes. Another theory that deals with time travel and causality is the "multiverse theory". This states that there are several parallel universes in our world and that every possible decision or action we could take takes place in another universe. So in this theory, a time traveler can go back in time and change events, but they would do so in a different universe and therefore it would have no effect on the universe they came from. This theory assumes that cause and effect are linked in every universe, but that the events in each universe are unique and therefore have no effect on other universes. So if a time traveler travels back in time and changes events, he would only affect the universe he is in. There are also theories that state that time travel is fundamentally impossible and that the idea of cause and effect in time is only based on a linear concept of time. These theories assume that the future is not predetermined and that every action we take influences our future. Assuming that time travel is impossible means that the past is unchangeable and that events in the future only happen because of decisions and actions in the present. In this idea of time, cause and effect are linked, but there is no influence from the future on the present or past.

Time travel and time travelers

How does time travel change the time traveler? One of the most obvious changes that can result from traveling through time is the experience of different eras and cultures. For example, a time traveler who travels to the Middle Ages would experience a completely different world than in the present day. This experience can broaden the time traveler's view of the world and help him or her to see their own culture and time from a new perspective. The time traveler may also develop a deeper appreciation for the history and development of humanity. Another change that can be brought about by time travel concerns the time traveler's knowledge and skills. By traveling back in time, the time traveler can, for example, learn the language or technologies of a certain era and thus expand their skills. Knowledge of historical events can also contribute to the time traveler having a more comprehensive understanding of the context and developments. But time travel can also have a psychological effect on the time traveler. For example, some time travelers may experience a sense of alienation or loss when they travel back in time and are unable to see their friends and family again. On the other hand, they might also experience a sense of loneliness and isolation if they travel to a future world where their friends and family no longer exist or recognize them. These psychological effects can vary depending on the personality and individual experience of the time traveler. Another potential change that can be caused by time travel is the ability to change the future. If a time traveler travels into the past and influences events there, this could lead to a change in the future. The time traveler could end up in an alternate reality where their decisions have different consequences than in their original timeline. This could result in the time traveler not being able to foresee the consequences of their actions and having to deal with the unexpected results. Another potential change that can be caused by time travel concerns the moral choices that the time traveler must make.

Time travel and time paradoxes

Is there a solution to the time paradoxes? A time paradox occurs when time travel leads to a self-contradiction. One of the most well-known forms of time paradox is the grandfather paradox: Imagine traveling back in time and killing your grandfather before your father or mother was born. If your grandfather was not born, how could you have been born to travel back in time? The grandfather paradox is an example of a causal paradox, where time travel leads to a contradiction in the cause-effect relationship. However, there are also other types of paradoxes, such as the information paradox or the ontological paradox. The information paradox refers to the question of what happens when you send information into the past. For example, if you travel back in time and patent an invention before it was discovered by the original inventor, who would own the patent? The ontological paradox occurs when an object or piece of information has no clear origin. An example of this is the story of Heinrich, who goes back in time and publishes Beethoven's ninth symphony as his own composition. Who then is the true creator of the ninth symphony? There is no single solution to all the paradoxes associated with the idea of time travel. However, most scientists and philosophers believe that if time travel were possible, it would not lead to a causal paradox. This view is based on the assumption that there would be a universal law that would prevent time travelers from changing events that have already happened. This assumption is supported by the Novikov self-consistency principle, which was developed by the Russian physicist Igor Novikov. The principle states that any action of a time traveler that changes the past would actually contribute to the past remaining as it was. In other words, you cannot change your actions because they are already part of history. Another solution to the grandfather paradox is the multi-world interpretation of quantum mechanics. This interpretation states that every decision you make leads to a bifurcation of the universe in which every possible future is realized. So if you were to travel back in time and kill your grandfather, this would only happen in a parallel world in which you do not exist. Nothing would change in your own world.

Time travel and evolution

How could time travel influence human evolution? What if time travel were actually possible? How could it influence human evolution? To answer this question, we must first look at the basics of human evolution. Evolution is the process by which species develop and change over time. These changes can be due to natural selection, mutations or genetic drift. The key to evolution is adaptation to changing environmental conditions. If we now add time travel to the equation, new possibilities open up. If we could travel into the past, we could change or influence events that affect the future. For example, a small intervention in the past could lead to a certain disease being eradicated or a certain technology being developed earlier. But what impact would that have on human evolution? Well, if we change events in the past, we could also change the environmental conditions that our ancestors were exposed to. This could cause selection factors to change and steer human evolution in a different direction. For example, let's say we travel back in time and prevent the outbreak of a certain disease that has a high mortality rate. This could result in those who are immune to the disease no longer having the evolutionary advantages they had before. This change could affect the genetics of the population and, in the long term, lead to human evolution moving in a different direction. Another possibility would be that we travel into the future and bring technologies with us that our ancestors did not have at that time. For example, if we travel back to the Stone Age and give our ancestors weapons or tools that they could not develop themselves, this would increase their chances of survival and steer evolution in a different direction. Of course, there are also negative effects of time travel on human evolution. If we travel back in time and change events, this could lead to a butterfly effect, where small changes have a big impact on the future. This could cause evolution to develop in unpredictable and potentially harmful directions. In any case, it is difficult to predict how time travel might affect human evolution.

Time travel and simultaneity

How does time travel relate to simultaneity? Time travel and simultaneity are two concepts that are closely linked. Both are fundamental elements of physics and philosophy and have implications for our understanding of space and time. In this article, we will explore the relationship between time travel and simultaneity and how they are viewed in physics and philosophy. Let's start with physics. An important aspect of time travel is that it goes hand in hand with the idea of time machines. The idea of a time machine is that an object can be transported from one point in the future or the past to another point in time. However, this requires a non-linear relationship between space and time. An example of this non-linear relationship can be found in Einstein's theory of special relativity. According to this theory, time is relative and depends on speed and gravity. An object moving at a high speed will perceive a slower time than an object at rest. Similarly, time will pass more slowly in a stronger gravitational field than in a weaker one. These effects are known as time dilation. The idea of time dilation is an important aspect of time travel as it implies that it is possible to travel into the future or past by utilizing time dilation. When traveling in a strong gravitational field or at a high speed, time will pass faster for an observer outside the gravitational field or at rest. This would mean that a person making such a journey would travel into the future or past when they return to the original gravitational field or speed. Another important aspect of time travel is the idea of simultaneity. In physics, there is no absolute simultaneity, only relative simultaneity. This means that time passes differently for different observers. What appears to be a simultaneous event for one observer may not be simultaneous for another. An example of this is the twin paradox. Imagine that there is a pair of twins who separate and fly in opposite directions. One of them stays on Earth while the other flies into space at a high speed and then turns around and returns to Earth. Due to time dilation, the flying twin will be younger than the twin remaining on Earth when it returns. The paradox is that for the flying twin the journey has only taken a few years, while for the twin remaining on earth many years have passed.

Time travel and time travel agents

How could time travel agents influence history? One of the most interesting questions surrounding time travel and time travel agents is how they might influence history. If time travel agents were able to travel back in time and change events, this could have a dramatic impact on the world as we know it. One way in which time travel agents could influence history is through the so-called butterfly effect. The butterfly effect states that even small changes in the past can lead to big changes in the future. For example, a time travel agent could accidentally kill a mouse that would have survived a snake bite in the past. This could cause the food chain to change and other species of animals to become extinct or multiply, which in turn has an impact on human society. Another way in which time travel agents could influence history is by altering historical events. For example, a time travel agent could try to solve a famous murder case by traveling back in time and convicting the perpetrator before he can commit the crime. This could lead to the story developing in a completely different direction, as the consequences of this crime might not have occurred. However, time travel agents could also be used to prevent historical events that turned out to be catastrophic for humanity. For example, a time travel agent could travel back in time and try to prevent the outbreak of a deadly plague before it can spread around the world. This could help save millions of lives and steer the development of human society in a different direction. However, it is important to remember that the effects of time travel and time travel agents on history can be unpredictable and potentially chaotic. Even small changes in the past could lead to dramatic changes in the future, and it is difficult to predict what impact these changes will have. Another problem with using time travel agents is that they may not be able to control the events they influence. For example, a time travel agent might try to influence a historical figure to make a better decision, but that person might end up making an even worse decision instead.

Time travel and time travel experiments

What if we could prove time travel experimentally? To answer these questions, we first need to look at the idea of time travel. Time travel refers to the idea that an individual or object can travel backwards or forwards in time. This is a fascinating idea that has inspired many people, but there are also many challenges and questions associated with it. One important issue is whether time travel is even possible. Based on current science and our understanding of the laws of nature, it seems unlikely that time travel will ever be possible. The laws of physics that govern our universe limit our ability to influence time. Nevertheless, there are some theoretical possibilities for time travel. The general theory of relativity states that space-time can be curved, which would theoretically make it possible to travel backwards or forwards in time. There are also ideas such as quantum entanglement that suggest that information could be sent back in time. But what happens if we could prove these theoretical possibilities experimentally? It would undoubtedly be a major breakthrough in physics and revolutionize our understanding of the universe. But it would also raise many ethical and practical questions. One important point is that time travel could be potentially dangerous. A traveler in time could inadvertently change the past or the future, with unpredictable consequences. It is also unclear how time travel would affect causality. For example, if a traveler in time travels back in time and changes an important historical moment, how would that affect the future? Another problem is that time travel could be very expensive and difficult. The technology needed to transport a person in time is not currently available and it is unclear whether it can ever be developed. Even if it were possible to make time travel possible, few people would be able to afford it.

However, there are also potential benefits of time travel. Experimental proof of time travel could help us to expand our understanding of physics and gain new insights into the laws of nature. It could also help us to better understand historical events and shape our future. An interesting example of this is the so-called "grandfather paradox". I will discuss this at a later date.

Time travel and the time machine in pop culture

How have pop culture and science fiction shaped the idea of time travel? Time travel and the time machine have always been a fascinating topic in pop culture and science fiction. In this article, we'll take a look at how pop culture and science fiction have shaped the idea of time travel and the impact this has had on our society. The idea of time travel has been around since ancient times, but it wasn't until the 19th century that H.G. Wells presented the idea of the time machine and thus founded the genre of science fiction. The time machine, which plays a central role in Wells' novel, allows the protagonist to travel into the future and discover a dystopian world there. This novel not only influenced science fiction literature, but also pop culture. Since then, time travel and the time machine have become a central theme in films, TV series and video games. Many of them have also drawn on H.G. Wells' novel to tell their own stories. One example of this is the film series "Back to the Future", which was directed by Robert Zemeckis in 1985. In this movie, a teenager named Marty McFly travels back in time with the help of a time machine developed by a mad scientist named Doc Brown and accidentally triggers a series of events that endanger his own existence. The film was a huge success and not only thrilled audiences, but also shaped the idea of time travel in pop culture. The DeLorean, the car that serves as a time machine in "Back to the Future", has become an iconic symbol and is often used as a synonym for time travel. Another example is the TV series "Doctor Who", which has been broadcast on the BBC since 1963. In this series, an alien time traveler named the Doctor travels through time and space using his spaceship, the TARDIS, to experience adventures and save planets. "Doctor Who" has not only gained a loyal fan base, but also a cultural significance. The Doctor, who has been portrayed by various actors, has become a symbol of tolerance, peace and the spirit of adventure. These examples show how pop culture and science fiction have shaped the idea of time travel. Time travel has become a fascinating and often recurring theme in pop culture, contributing not only to entertainment value but also to reflection on the human condition. The idea of time travel has also had an impact on our society. It has contributed to our awareness of time and the importance of history. Time travel can help us understand the past and shape the future.

Time travel and quantum mechanics

How can we explain time travel through quantum mechanics? Quantum mechanics is a branch of physics that deals with the description of nature on the smallest scale. The theory of quantum mechanics states that particles do not behave like classical objects, but have the properties of both particles and waves. This wave function describes the probability of finding a particle at a certain place and at a certain time. One of the interesting consequences of quantum mechanics is the phenomenon of quantum entanglement. If two particles are entangled with each other, this means that the properties of one particle are correlated with the properties of the other particle, regardless of the distance between the two particles. Quantum mechanics allows particles to travel backwards in time. This is known as retrocausal influences. The mechanism that makes this possible is based on quantum entanglement and the fact that time in quantum mechanics is not regarded as an invariant constant, but as a variable quantity that depends on observation. The idea that quantum entanglement and retrocausal influences can enable time travel was first proposed in the 1990s by physicist David Deutsch. Deutsch suggested that it might be possible to go back in time by creating quantum entanglement between two particles and then sending one of the two particles back in time. This idea was later developed further by other physicists such as Seth Lloyd and Andrew Jordan. They proposed that time travel through quantum entanglement and retrocausal influences could not only be possible, but also consistent, as the laws of quantum mechanics would allow events in the past and future to be connected. An interesting example of how quantum mechanics and time travel could be related is the famous "Grover's Algorithm". This algorithm, developed by Lov Grover in 1996, uses quantum entanglement to search faster than classical algorithms. One way to interpret Grover's algorithm is that it performs a kind of time travel. The algorithm searches through all possible states of the system simultaneously and thus finds the correct solution much faster than a classical algorithm. This is an example of how quantum mechanics and time travel can be linked in a way that allows us to perform faster calculations.

Time travel and the time travel paradox

What is the time travel paradox? The time travel paradox is one of the most intriguing and confusing concepts in science fiction literature and movies. It describes the potential consequences that can occur when someone travels back in time and changes the past. The paradox arises when the changes to the past affect the future, which can lead to the original cause of the time travel no longer existing or having evolved in a completely different direction. To better understand the paradox, we can imagine a simple example: A man named John travels back in time to 1955 to meet his parents before they met and married. If John now successfully changes the past, for example by preventing his parents from meeting, then this would mean that John himself was never born. But if John was never born, then he can't travel back in time to meet his parents and so the reason for his time travel no longer exists. Another example that is often used in films and books is the so-called grandfather paradox. A man travels back in time to kill his grandfather before his father was born. If he does this successfully, his father would never be born, which would mean that he himself was never born either. This in turn would mean that the man could never travel back in time and kill his grandfather. The time travel paradox raises a variety of questions that are not easily answered. For example: What happens to the timeline when someone makes changes in the past? Is the future changed? Or are there alternative timelines that are created when someone travels into the past? Is it even possible to change something in the past without destroying the entire universe? There are various theories that attempt to explain the time travel paradox. One of these theories states that the universe reacts in a way that prevents anyone from changing anything in the past. For example, someone trying to kill their grandfather could be repeatedly prevented from doing so by chance, or it could be that their grandfather miraculously survives in each scenario. Another theory is that alternate timelines are created when someone travels back in time and makes changes. In this theory, the original timeline would continue to exist while a new timeline is created that is influenced by the changes to the past. In this scenario, the time traveler would never change their own timeline, but merely create an alternate timeline.

Time travel and the space-time continuum

How are space and time related? Space and time are two fundamental concepts that are closely linked in physics. They form the spacetime continuum, which determines our reality and influences our understanding of the world around us. In the following, we will gain an insight into the complex connection between space and time and how time travel fits into this concept. The spacetime continuum The spacetime continuum is a mathematical model that describes the physical world. It combines the three dimensions of space with the fourth dimension of time to form a four-dimensional continuum. The concept of the spacetime continuum was first introduced by the German physicist Hermann Minkowski, who extended Albert Einstein's theory of special relativity. The connection between space and time The connection between space and time is a fundamental concept of modern physics. According to the theory of special relativity, the speed at which an object moves is relative to the position of observation. This theory also states that space and time are intertwined, and that the gravity we experience in the world around us is caused by the curvature of the spacetime continuum. Another concept that illustrates the connection between space and time is Lorentz contraction. This theory states that objects moving at a high speed are shortened longitudinally due to the change in spacetime distances. This means that space and time are connected in such a way that they can influence and change each other. Time travel The concept of time travel refers to the possibility of traveling in time and visiting events that have already happened or have not yet happened. Although time travel is not possible in reality, there are many fascinating thought experiments that give us an insight into the possible effects of time travel on the space-time continuum. The possibility of time travel is often explored in physics through the theory of general relativity. One of the main implications of this theory is that the curvature of the spacetime continuum allows for the possibility of time travel. A black hole, for example, is an object with such a high density and such a strong gravitational force that it can bend nearby spacetime and even affect time. Another concept associated with time travel is the grandfather paradox. This paradox states that if someone were to travel back in time and kill their own grandparents, they themselves would never exist and thus the possibility of committing the act would not exist. This shows that time travel can potentially lead to inexplicable paradoxes.

Time travel and the timeline

How does time travel change the timeline? In this article, we will take a closer look at how time travel affects the timeline and the impact this can have on the world. To understand how time travel affects the timeline, we first need to understand the basics of the timeline. The timeline is essentially a sequence of events that take place in the past, present and future. Each event on the timeline has a specific cause and effect, which can be summarized in a chain of causality. When we perform an action, it triggers a reaction that has an effect on the timeline. This means that every change on the timeline has an effect on all other events. If we now travel back in time and intervene in the past, we have the opportunity to change the chain of causality. A small intervention in the past can have an impact on the future and change the timeline. For example, a person intervening in the past and killing a butterfly could cause a storm to be triggered in the future with catastrophic effects. This concept is known as the 'butterfly effect' and illustrates how small changes in the past can have a big impact on the future. Changing the timeline through time travel also has an impact on people's identities. When we travel into the past and change our own past, we also affect our own identity. We could erase ourselves from existence or become a different person than we originally were. These concepts have been portrayed in numerous movies and books, such as "Back to the Future" or "The Time Machine Manifesto". There is also the idea that there may be alternative timelines created by time travel. If we travel back in time and change something, an alternative timeline may be created that is different from the original one. This means that we could be living in an alternate reality based on a change in the past, while the original reality continues to exist. This concept is known as the "multiverse" and is a common motif in science fiction literature and film. It is important to note that time travel is only a hypothetical concept and has not been proven.

Time travel and the time travel paradox

How can we resolve the time travel paradox? However, the possibility of time travel also brings with it some paradoxes that have not yet been fully resolved. In this essay, we will look at the time travel paradox and possible solutions. The time travel paradox occurs when a person travels back in time and changes events that lead to a change in the present. If the present is changed, the person would have no need to travel to the past as the events would have happened differently. This leads to a contradiction as the person must have been in the past to change the events, but if they do, this may result in them never having traveled to the past. There are several variants of the time travel paradox, such as the grandfather paradox or the information paradox. One way to resolve the time travel paradox is to assume a multiverse theory. The theory states that every time a decision is made, a new universe is created in which the alternative decision is made. If someone travels back in time and changes events, this would create a new universe that is different from the original universe. In this new universe, the person would have intervened in the past, but it would have no effect on the original universe where the person did not intervene.

Another possibility is that time travel has already taken place in the past and that the events we experience today are the result of time travel in the past. This theory states that every action we take has already been considered by time travelers in the past. If someone travels back in time to change events, he/she will actually have already been in the past and influenced events to lead to the present we experience today. Another possibility is the idea that time is unchanging and that every action someone takes in the past is already part of the original timeline. If someone travels back in time to change events, this would actually result in the events taking place exactly as they have already happened. This is known as the 'consistency principle' and is a popular theory in many time travel stories.

Time travel and the future

What can we learn from the future? The idea of time travel has long captured the human imagination. The possibility of traveling to the past or the future has fascinated people for centuries. Although there is as yet no scientifically proven way to actually travel through time, we can still learn valuable lessons from imagining and speculating about the future. A look into the past shows us how the world has changed over time. We can learn from history and make our decisions based on what has gone well and badly in the past. If we look at the technology of the last few centuries, we can see how quickly things can change. A few decades ago, the idea of a mobile phone was unthinkable, but today we have smartphones that keep us constantly connected to the world. It's amazing to see how quickly the world has changed in terms of technology, science and business. So if we take a look into the future, we can prepare ourselves for what's to come. One of the most important lessons we can learn from imagining and speculating about the future is the importance of preparation. If we prepare for possible future developments, we can better adapt and respond. One example of this is the climate crisis. Although there was no evidence of climate change in the past, we now have enough data and scientific evidence to know that we need to act. By preparing for the future, we can prepare for potential challenges and develop solutions. Another important lesson we can learn from imagining and speculating about the future is the importance of collaboration. The challenges we will face in the future are often global in nature and require cooperation on an international level. The climate crisis is an example of this. To overcome this challenge, we need to work together and find global solutions. If we look at history, we see that many of the biggest challenges humanity has faced have been overcome through cooperation and joint efforts. Another important aspect of imagining and speculating about the future is the importance of vision and ideas. New technologies, scientific discoveries and cultural changes are often driven by visionaries and innovators. Through their ideas and imaginations, they can shape the future and break new ground. History shows that visionaries such as Leonardo da Vinci, Isaac Newton and Albert Einstein have changed the world by developing new ideas and concepts. Even if the idea of time travel remains just a mind game, we can still learn valuable lessons from imagining and speculating about the future.

Time travel and infinite time

What is infinite time? The idea of time travel and infinite time has always fascinated mankind. What is time and how does it work? How far can you travel into the past or the future? Is time infinite and if so, what does that mean for us as a species? Infinite time is a concept that is often beyond human imagination. It is difficult to imagine that there is a beginning or an end to time. When we speak of infinite time, we are referring to a reality in which time is not bound to a specific event or time. Time simply exists as a continuum that is infinitely extended in both directions. When we deal with infinite time, we have to deal with the concepts of space and time. Modern physics has shown that space and time do not exist independently of each other, but function together as spacetime. This means that space and time behave in a way that is intertwined. A change in space therefore also leads to a change in time. In science, the concept of spacetime is described by the theory of relativity, which was developed by Albert Einstein. This theory states that time is not constant, but changes depending on speed or gravity. An event that takes place in one space can take place in another space at a different time. This is known as the concept of time dilation. When we consider time travel, we also have to deal with the idea of causality. Causality states that an event has a cause and this cause in turn leads to another event. Generally, events follow a linear sequence in which each event follows the previous one. However, time travel could change this idea of causality. Traveling back in time and taking an action that changes the past could lead to a change in the future. This would mean that events no longer necessarily occur in a linear sequence, but that the future could be influenced by the past. Another interesting concept relating to time travel and infinite time is the grandmother's paradox. This paradox asks the question of what would happen if you traveled back in time and killed your grandmother before she had children. If you kill your grandmother before she has children, it would mean that you yourself would never be born. But if you are never born, who killed your grandmother? This paradox shows how difficult it can be to reconcile the idea of time travel and causality. There are also many popular cultural representations of time travel and infinite time. Films such as "Back to the Future" and "Interstellar"

Time travel and the Big Bang

How are time travel and the Big Bang connected? Time travel and the Big Bang are two concepts that are not necessarily associated with each other at first glance. But a closer look reveals a connection between them. The Big Bang is the starting point of the universe, which created time, space and matter. It occurred around 13.8 billion years ago and marks the beginning of our cosmic history. But what if it were possible to travel back in time and observe the Big Bang? The idea of time travel is fascinating and has long been the subject of science fiction films and books. But how realistic is it? Modern physics has taught us that time and space are interconnected and that they behave in unimaginably complex ways. According to Einstein's theory of relativity, time passes more slowly near masses, and if you travel faster than the speed of light, you could theoretically travel into the future. But what about the past? Could we actually travel back and experience the Big Bang? The answer to this is more complicated than you might think at first glance. Because in order to observe the Big Bang, we would have to go to the place where it happened - and that is more difficult than expected. The Big Bang occurred at a point in space that was so tiny that it was smaller than an atom. There was therefore no place where the Big Bang could be observed. Furthermore, in the first few moments after the Big Bang, the universe was extremely hot and dense, so that no matter, radiation or light could escape. Only when the universe expanded and cooled down were the first elementary particles and atoms formed that could emit light and radiation. So if you want to observe the Big Bang, you would have to reproduce the conditions of the early universe - an undertaking that is currently only possible in particle accelerators such as the Large Hadron Collider. Here, protons are shot at each other with extremely high energy to create conditions similar to those in the early universe. But even this only gives us limited insights into the early stages of the universe. But even if it were possible to observe the Big Bang, the question would still remain as to whether we were traveling back in time or just to a different place in the universe. This is because time travel is associated with numerous paradoxes and logical contradictions that have not yet been fully resolved. One example of such a paradox is the grandfather paradox: if you travel to the past and kill your grandfather before he can father children, how is it possible for you to exist in order to make the journey?

Time travel and the future of humanity

How might the future of humanity be affected by time travel? First of all, it is important to understand that time travel as we know it from movies is highly unlikely. The idea of traveling back in time and changing events is a type of time travel known as a "time loop" and is considered improbable in physics. The idea of traveling to the future is also theoretically possible, but there is no known way to change the past. However, time travel could influence the future of humanity in other ways. One possibility would be that a future civilization develops some sort of time travel technology and travels back to our time to influence us. This idea is often referred to as "Temporal Intervention". An example of a possible Temporal Intervention would be that a future civilization travels back to our time to warn us of a catastrophe that we would otherwise not survive. This could lead to us taking action to avert the catastrophe and thus change the future of humanity. Another possibility would be that a future civilization travels back to our time to help us solve a particular problem. For example, they could provide us with a technology or concept that helps us solve a global problem such as climate change. It is also conceivable that a future civilization could travel back to our time to save us from ourselves. For example, if we develop a technology that enables us to destroy ourselves, a future civilization could travel back and prevent us from developing that technology. However, there are also possible negative effects of time travel on the future of humanity. If a future civilization travels back to our time and tries to influence us, it could lead to unforeseen consequences. For example, their actions could cause history to unfold in a completely different way than it otherwise would have. Furthermore, a future civilization that masters time travel could also abuse power. They could try to manipulate the past in their favor to increase their own power and influence. This could lead to a dystopian future in which people's freedom and rights are suppressed. It is also possible that time travel could lead to a kind of race between different future civilizations.

Time travel and the time travel experiment

What could a successful time travel experiment look like? Time travel has long been a fascinating topic in science fiction and has also sparked interest in the scientific community. The idea of traveling through time and experiencing the past or the future has always fascinated mankind. A successful time travel experiment would mean that it is possible for us to travel back or forward in time and not change the future just by going back in time. What could such an experiment look like? A successful time travel experiment requires both a deep understanding of the nature of time and advanced technology. Albert Einstein's theory of special and general relativity provides a mathematical basis for time travel, stating that time is a relative concept influenced by gravity and velocity. The theory suggests that the faster you move or the stronger the gravity you experience, the slower time passes. So to conduct a successful time travel experiment, we would need to find a way to move our consciousness or body through time and space without affecting the environment. One possibility would be to build a spaceship that can fly at a speed close to the speed of light to experience the effect of time dilation. Another possibility would be to build some kind of time machine that allows us to travel to the past or future without changing the environment. Most theories about time machines involve the use of wormholes that would allow us to travel through the space-time continuum. Such a wormhole would allow us to establish a connection between two distant places in space and time. To conduct such an experiment, we would first need to find or create a wormhole. One approach would be to create some kind of singularity, a point where gravity is so strong that spacetime collapses. This would create a wormhole that could transport us through space and time. Once found or created, we would need to build a spaceship or capsule that could transport us through the wormhole. The capsule would have to be constructed in such a way that it could experience the effect of time dilation and that when traveling into the past or future, it would not alter the environment. The capsule would also have to be designed to survive the journey through the wormhole undamaged. To conduct the time travel experiment, we would also need to develop a method to determine whether we have actually traveled through time and whether we have actually had no effect on the environment.

The dream of time travel

The idea of time travel is intriguing as it would allow us to correct our past or influence our future. We could travel back in time to relive important historical events or to change the world by making important decisions differently. We could also travel into the future to see how our world has developed in a hundred or a thousand years. In science, the topic of time travel is also discussed time and again. There are theories that say it would be entirely possible to go back in time or travel into the future. Some scientists believe that it would be possible to travel back in time using black holes or wormholes. However, there are also many counter-arguments and difficulties in implementing these theories. Although the idea of time travel sounds appealing, there are many complex physical and mathematical problems that need to be solved. In addition to the scientific aspects, there are also many ethical and moral issues surrounding time travel. If we travel back in time and change the past, what would that mean for the present and the future? Would we create a new reality or change an existing reality? Would we perhaps even destroy the future? The idea of time travel also has an impact on our concept of time and space. If we could travel in time, wouldn't we also move in space? Could we also travel through the universe in this way and explore other planets or galaxies? The fascination with time travel also has an impact on popular culture. Stories dealing with time travel are told time and again in literature and film. Well-known examples include "Back to the Future", "Doctor Who" and "Timeless". These stories not only show us what time travel could look like, but also what impact it could have on our world. All in all, time travel remains a fascinating topic that will always keep us busy.

Time travel and teleportation

Let's start with time travel. Time travel is essentially a journey into the past or the future. The idea of time travel has always been an integral part of the human imagination and appears in numerous myths and legends. In modern scientific fiction, however, the idea has been further developed through the influence of the theory of relativity and quantum mechanics. One of the most important concepts in time travel is the idea of spacetime. Spacetime is a physical concept that considers space and time as a single entity. When we move in spacetime, we move in a four-dimensional world. When we move in a straight line through space, we are also moving through time. As a result, time is different for an observer moving relative to us than it is for us. This idea is known as "time dilation". There are various theories about how time travel could be possible. One possibility would be to use a "wormhole" or an "Einstein-Rosen bridge". A wormhole is essentially a connection between two points in the space-time continuum that would allow a traveler to get from one point to another without traveling through the space in between. It is believed that a wormhole could also be used as a type of "time tunnel" that would allow a traveler to travel into the past or the future. Another possibility for time travel is the use of black holes. Black holes are extremely dense objects that generate such a strong gravitational pull that nothing, not even light, can escape them. It is believed that a traveler who travels into a black hole could be pulled back in time due to the strong gravity. It is important to emphasize that these concepts are currently only theoretical and that there are no practical ways to undertake time travel. There are also numerous paradoxes and logical problems associated with the idea of time travel, such as the grandfather paradox. This paradox states that if someone travels back in time and tries to kill their own grandfather, they would never have been born, which then means they could never have killed their grandfather.

Black holes

Time travel and black holes are two fascinating topics that have long captured the human imagination. While the idea of traveling in time has been discussed in literature and philosophy for centuries, the existence of black holes has only been confirmed by astronomical observations in recent decades. But how are these two concepts related and what can we learn from them? A black hole is formed when a massive star collapses at the end of its life and reaches such a high density that even light can no longer escape. Black holes have an incredible gravitational force and can even influence entire galaxies. But what does this have to do with time travel? The idea that you could travel through a black hole to another time or dimension was first proposed by the British physicist Stephen Hawking. He postulated that black holes emit energy known as Hawking radiation. This radiation causes the black hole to slowly shrink and eventually evaporate. Hawking speculated that this radiation also contains information about the contents of the black hole, including information about everything that has ever fallen into the black hole. The idea that this information could also enable time travel was further developed by other physicists. They argued that if you dropped an object into the black hole and later measured the Hawking radiation, you would be able to collect information about the object that had been lost in the past. This information could then be used to return the object to the past. Of course, the idea that you could travel back in time through a black hole is still very speculative. There are many unknowns and hypotheses in theoretical physics that make it difficult to make such predictions. However, it is certain that black holes will expand our understanding of space and time. Another interesting idea associated with black holes and time travel is the possibility of traveling to the future through time travel. A well-known example of this is the "twin paradox". It assumes that a twin gets into a spaceship and flies into space at almost the speed of light, while his brother remains on Earth. When the flying twin returns, he will find that he has aged much more slowly than his brother on earth. This idea is not just science fiction, but has actually been proven experimentally. The effect is explained by relativistic time dilation, which states that time passes more slowly for an observer in a moving system than for an observer in a stationary system.

The grandfather paradox

is one of the best known and most discussed paradoxes in the context of time travel. It refers to the idea that someone traveling through time may inadvertently change events in the past that could have an impact on the future. In this case, the question arises as to whether the changes to the past would have an effect on the person themselves who undertook the journey. The consequences of these paradoxical situations are often described as impossible or contradictory. The grandfather paradox is named after the hypothetical situation in which someone travels through time and attempts to kill their own grandfather before he has fathered their father. This hypothetical situation has often been used as an example of the paradox, but there are many other similar situations that have also been discussed. The paradox is that if someone travels through time and performs an action in the past that changes the events of the future, the consequences of that change could have a retroactive effect on the action the person took. In other words: If someone travels through time and performs an action in the past that results in themselves not being born, how can they even be in the past to perform that action? Another example of the grandfather paradox is the idea of someone traveling through time and preventing an event from occurring in the past that caused them to make a certain decision in the future. If he changes the past, how can he make the decision in the future that he would have made if the event in the past had not been changed? One possible solution to the grandfather paradox is that each change to the past creates an alternative timeline in which the changes took place, while the original timeline remains unchanged. This is also known as the many-worlds interpretation of quantum mechanics. According to this interpretation, there is an infinite number of parallel worlds that exist at every moment of time, and every decision that is made creates a new reality. Another possible solution is that the universe has a mechanism to prevent paradoxical events. For example, the universe might be able to influence events in the future so that changes in the past have no effect on the future. There is also the possibility that the grandfather paradox is simply unresolvable. It could be that the idea of time travel is paradoxical and that there is no way to resolve all the contradictions involved.

Time travel and determinism

But how does the concept of time travel relate to the concept of determinism? One of the main issues when discussing time travel and determinism is the question of causality. Causality refers to the fact that an action or event has a cause, which in turn has an effect on something else. If you travel back in time and change something in the past, this could have an effect on the future and therefore change the course of events. However, if everything is predetermined, as determinism states, then you can't really change anything, as everything that happens in the future is predetermined based on present circumstances. One way to solve this paradox is the idea of the timeline. A timeline is a concept that states that there are multiple parallel universes, each with its own reality and history. Traveling back in time and changing something opens up a new timeline that is different from the original one. This would mean that the future you have visited is predetermined, but that there are also alternative realities that are created by changing the past. Another concept that plays a role in the discussion of time travel and determinism is the concept of predestination. Predestination refers to the idea that everything that happens in the future is already predetermined and that there is no free choice. This would mean that every action you take is already predetermined and that you have no influence on the future. If you combine predestination with the idea of time travel, this would mean that every action you take in the past is already predetermined and that you cannot change the future. This would mean that the future you have visited will always remain the same, regardless of what actions you take.

The probability of time travel

But is time travel really possible? And if so, how likely is it that we will one day be able to travel through time? First of all, we have to ask ourselves what time actually is. For us humans, time is a kind of yardstick with which we measure the changes in our universe. But in physics, time is much more than that. Time is a physical dimension, similar to space. It is part of the space-time continuum described by the general theory of relativity. When we talk about time travel, we also have to deal with Einstein's famous equation $E=mc^2$. This equation states that energy and mass are the same thing. In other words, energy can be converted into matter, and matter can be converted into energy. This is why science fiction is so often about time travel - because time travel is linked to energy and space. Another concept that is important when we talk about time travel is that of singularities. A singularity is a point in the space-time continuum where the laws of physics break down and time ceases to exist. An example of a singularity is a black hole. Inside a black hole there is a point where the density is infinite and the gravity is so strong that even light cannot escape.

Now, when we talk about time travel, we have to ask ourselves whether it is possible to use singularities to travel through time. One approach that is being discussed in science is the idea of wormholes. A wormhole is a connection between two points in the spacetime continuum that could make it possible to travel through time. However, there are also some challenges in using wormholes for time travel. For one thing, wormholes are unstable and could collapse at any time. For another, the extreme gravitational forces near a wormhole would likely tear a human or spaceship apart. Another way to make time travel possible would be to use time dilation. Time dilation occurs when an object is moving at a high speed. This was described by Einstein in his special theory of relativity. Time appears to pass more slowly the closer you get to the speed of light. This could theoretically be used to travel into the future. For example, if you could build a spaceship that moves at high speed, time would pass more slowly on board the spaceship than on Earth.

Imprint

Tom Held
Am Anger 3
06869 Coswig
Germany
Luna-Publishing.de

www.ingramcontent.com/pod-product-compliance
Lightning Source LLC
Chambersburg PA
CBHW071017290526
45795CB00005B/1845